Sleep and Rest in Animals

Corine Lacrampe

Sleep and Rest in Animals

FIREFLY BOOKS

A Firefly Book

Published by Firefly Books Ltd., 2003

First Printing

National Library of Canada Cataloguing in Publication Data
Lacrampe, Corine
Sleep and rest in animals / Corine Lacrampe.
Translation of : Dormir, rêver : le sommeil des animaux.
Includes bibliographical references and index.
ISBN 1-55297-677-7
1. Sleep behavior in animals—Juvenile literature. I. Title.
QL755.3.L3213 2003   j591.5'19 C2002-903958-4

Publisher Cataloging-in-Publication Data (U.S.)
Lacrampe, Corine.
    Sleep and rest in animals / Corine Lacrampe.—1st ed.
[112] p. : col. ill. , col. photos. ; cm.
Includes bibliographical references and index.
Summary: An investigation of the sleep habits of insects, reptiles, amphibians, birds and mammals.
ISBN 1-55297-677-7 (pbk.)
1. Sleep behavior in animals. 2. Animal behavior. I. Title.
591.519 21   QL755.3L33 2003

Published in Canada in 2003 by
Firefly Books Ltd.
3680 Victoria Park Avenue
Toronto, Ontario M2H 3K1

Published in the United States in 2003 by
Firefly Books (U.S.) Inc.
P.O. Box 1338, Ellicott Station
Buffalo, New York 14205

Printed in Italy

Acknowledgments

None of this would have been possible without the approval and support of Mireille Cayreyre. Special thanks go to Jean Dunia, not only for his constant, precious help but also for his encouragement and his confidence in the project. Lastly this book owes much to Emmanuelle Vetard's help as well as to Philippe Ement's stringent proofreading.
S. de S.

Special thanks go to:
Anne Hauben, chief editor of Science et nature, for her confidence and friendship.
For their scientific contribution, thanks also go to:
Irène Tobler, Professor of biology, research scientist specializing in comparative studies on sleep (Pharmacology and toxicology laboratory at Zurich University, Switzerland).
Jean-Louis Valatx, Director of research at Inserm, in Lyon, France
Gérard Dewasmes, sleep physiologist, CNRS research scientist (Medical School, Amiens, France).
Bernard Buisson, Alain Blanc and Joël Attia, research scientists at the science faculty, Saint-Etienne, France (work on the snail's body clock)
Gilles Bœuf, Director of the Arago laboratory, Banyuls-sur-mer, France professor of fish physiology, Paris VI University, France.
Claire and Jean-François Voisin, ornithologists, lecturers at the Museum of Natural History in Paris, France.
Emmanuelle Grundmann, orangutan specialist (Museum of Natural History in Paris, France).
C. L.

Many thanks to Adrian, Alix, Pip and Yves for their patience!
B. P.

Editorial Director:	Sophie de Sivry
Editor and iconography:	Valérie Millet
Translation:	Bridget Poher
Scientific collaboration:	Henry Depoortere
Reader for scientific terms:	Henry Depoortere, David Sanger
Design and layout:	Double, Paris
Photoengraving:	Arciel Graphic, Paris

Contents

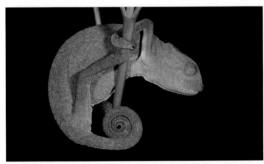

Foreword

Do all animals sleep? Although researchers have unraveled some of the mystery of human sleep, they have examined only about 200 other species. All animals, from bees to elephants, switch between phases of activity and inactivity. Birds and mammals are endothermic (warm-blooded, generating their own internal heat) and undergo repetitive stages of sleep, with corresponding changes in electrical activity in the brain. The first stages of sleep show slow waves and the last stage exhibits agitated waves—similar to the awake state—and indicate the animal is dreaming. Ectothermic animals (cold-blooded, depending on the environment to warm up or cool down) such as arthropods, fish, amphibians and reptiles have different nerve or brain structures. They produce electrical activity from nerve centers or brains but do not show the same kinds of changes in brain activity as can be seen in mammals and birds.

It is difficult to trace the evolution of sleep because behavior does not leave a fossil record. As the brain and nervous system of certain animals grew more complex during evolution, the sleep process became increasingly complicated.

Sleep and rest are manifested in extremely diversified ways in animals. We must observe daily and seasonal behavior and changes that occur in states of hibernation, estivation, dormancy or torpor, and diapause. Simply put, hibernation occurs in cold climates: the animal seeks out shelter and enters a deep sleep with its heart rate, body temperature and activities drastically reduced. Estivation is the same process but in hot, desiccating climates where the animal "sleeps" until the rains come. In dormancy and torpor, some animals enter a lethargic state with a slightly reduced metabolism; it could be as short as a single night (a hummingbird) or as long as the winter (a bear). Diapause is a long, dormant period when all development ceases.

Interest in sleep is not new. Ancient Sanskrit writings refer to the brain's various levels of wakefulness: alertness, dreamless sleep and dreaming. Early observers had intuitively noted the different characteristics of sleep in animals. In the first century A.D., Lucretius, the Roman poet and philosopher, observed that horses and dogs dream because they continue to display physical activity while sleeping. In the fourth century B.C., in his work *De somno et vigilia* (On sleeping and waking), Aristotle claimed that digestion produced sleep. Warm vapors brought on by digestion rose to the brain to be cooled before descending to cool the heart, which he considered the body's sensory center.

According to Greek mythology, Hypnos, the god of sleep, regulates the rhythm of our lives while his brother Thanatos is god of death. What differentiates sleep from wakefulness or death? Medhi Tafti, a neurophysiologist at the University of Geneva points out that the first behavioral indicator in sleep is lack of movement. However, he stresses that many animals spend much of their time in a calm state where little movement takes place even if they are not sleeping. On the other hand, dolphins move constantly while sleeping. The sleeping state must also be instantly reversible if it is to be differentiated from death or hibernation.

Irene Tobler, professor of biology at the Institute for Pharmacology and Toxicology at the University of Zurich in Switzerland, is one of the leading specialists on laboratory studies of sleep in animals. Professor Tobler has set out six main sleep indicators: the animal selects a particular spot for sleeping; it adopts a typical sleep posture; it is in a quiet, relaxed physical state; its waking threshold is high; it rapidly passes from being asleep to being awake; and if an animal is deprived of the state that resembles sleep for sufficiently long enough, it will recuperate later by sleeping.

The invention of the electroencephalogram (EEG) revolutionized the scientific study of sleep. In 1875, Richard Caton, an English scientist, placed electrodes on a rabbit's brain to record electrical activity. Using a galvanometer, he was able to measure the frequency of EEG waves. In 1924, Hans Berger, a German doctor, recorded the electrical activity of the human brain. From then on, researchers carried out their measurements on anesthetized animals, particularly cats and rabbits. Klaue, another German researcher was the first to make a recording of the brainwaves of a conscious animal. In 1937 he published an article which described a cat's light sleep with slow (delta) waves followed by deep sleep with rapid cortical activity. He had, in fact, discovered paradoxical or REM sleep.

The first scientist to devote himself to sleep research in the 20th century was Nathaniel Kleitman, who published *Sleep and Wakefulness* in 1939 and established the world's first sleep research laboratory at the University of Chicago. In 1953, Kleitman and Eugene Aserinsky, one of his students, revolutionized sleep research with their article in the journal *Science* on their discovery of REM (Rapid Eye Movement) sleep in humans, a type of sleep whereby rapid ocular movements are perceptible under the eyelids. By spending many nights recording EEGs on volunteer sleepers, Aserinsky and Kleitman established an indisputable link between dreaming and REM sleep. When awakened during the REM phase, more than 80% of patients said they were dreaming.

Also spearheading research on sleep in his Lyon laboratory was French neurophysiologist Michel Jouvet. He carried out studies on sleep in cats in the 1960s and succeeded in isolating the area of the brain responsible for the muscle paralysis that accompanies REM sleep. Although the body is effectively paralyzed in REM or paradoxical sleep, cerebral activity is similar to that of wakefulness, thus creating a paradox. Jouvet disabled the nerve impulses that paralyzed the cat's muscle activity during REM sleep and observed the dreaming cat going through various postures of attack and defense.

In similar fashion to Dr. Tabler, Michel Jouvet considers sleep to be a regular phenomenon indicated by lack of movement, a decrease in cardiac and respiratory rhythms, a reduction in electrical activity in the brain (in animals with a brain), and an increase in waking thresholds.

In this book, we will stretch the meaning of the word "sleep" and suspend some of the sleep-defining criteria for certain animals as we learn about birds that sleep while flying, fish and marine mammals that sleep while swimming, and frogs that survive a death-like freezing state.

1
Do insects sleep?

WE USUALLY SEE INSECTS CRAWLING or flying or scuttling about—the cockroach eluding our stomp, the mosquito sucking up our blood, the butterfly gracing our gardens. We consider ants and bees symbols of hard work. Nevertheless, an insect's life usually contains periods of rest, although some species have a very short, frenetic adult life.

Arthropods (an animal with an exoskeleton, segmented body and paired and jointed legs) represent the largest group of animals in the world. Insects are only one class of arthropod. In addition to insects, this group is made up of arachnids, millipedes, centipedes, and crustaceans.

Although we cannot technically say that insects sleep, the beautifully metallic cuckoo or jewel wasp (family Chrysididae, photo on left) is definitely resting with antennae tucked under its head, wings folded down, head bent, motionless and quiet. During this inactivity, there is a reduction in the electrical activity in nerve centers that serve as the insect's brain but this change in electrical activity is not comparable to the mammalian stages of sleep waves.

The larva of the bagworm moth (family Psychidae, photo on right) never comes out of its protective case. As soon as it hatches from an egg, it begins to make its bag out of spun silk, leaves and other plant material. When it wants to feed, it sticks its head out; when it wants to rest, it withdraws into the bag. It pupates within the bag and has a short adult life, programmed solely to reproduce the next generation. It is difficult to say if this insect sleeps.

Log cabin
Above is the aquatic larva of the caddisfly (Limnephilidae) in its tubular case, which it makes from sticks, pebbles and other debris. Some use large twigs to prevent fish from swallowing them. This casing provides shelter and camouflage before the larva's final molt into a winged adult.

15

Early evolution

We don't usually see the millions of minute insects that are around us, hidden under rocks or behind bark, in the ground or under water. Insects appeared about 400 million years ago, long before the dinosaurs, and have always been extraordinarily diversified. Entomologists estimate there are between one and five million species of insects on our planet. They live on land, in the air, and in water and have colonized almost every type of habitat. Although many humans dismiss or loathe them, insects are indispensable to the ecological balance of our planet. In addition to being an essential part of the food chain, they pollinate flowers and as decomposers take care of biological recycling.

Day and night shifts

During the day moths use camouflage for protection while resting on a tree, blending in with the muted colors of the bark (see pages 8–9 and 16). If you touch one gently, it will remain in its somnolent state and will only "awaken" and fly away if you use more force to dislodge it from its place. Clearly we can loosely say the moth is sleeping. The leaf insects (see pages 10–11) and their cousins, the walkingsticks (see pages 18–19), use mimicry and look just like the leaves and twigs where they rest. At night they will become active and feed on those same leaves because the danger of bird predators is over until the sun rises again.

PERFECT MIMICS

The walkingstick or phasmid may hold the record for resting motionless. It imitates a twig and will not move during the day when bird predators might catch it, but it becomes active at night to feed. It will sway in the breeze to match the bending of nearby plants, and if it gets blown to the ground, it will play dead until it is safe to get up under cover of darkness. When a stick-insect moves, it moves very slowly. Some species change color, becoming lighter by day and darker by night, triggered by the circadian rhythm of light and dark. Some have wings but most are wingless.

Cockroaches actually sleep 14 hours a day but leave their hideaways at night to forage in the dark. On their way, silverfish cross paths with the cockroaches before sliding behind the wallpaper to digest some cellulose. In the yard, from vegetable patch to treetops, everything is active in the moonlight. Moths sip nectar, crickets sing their songs, the mole crickets dig up other insects' larvae. As the night ends with the first rays of the sun, the nocturnal inhabitants of this swarming world take shelter to rest or sleep, and the day shift of diurnal insects takes their place.

Sleep postures

The resting posture of invertebrates is not sufficient to indicate sleep because many of them rest while still being alert to danger. They would never have survived so successfully all these millions of years if they had not adopted survival tactics that leave them always partially alert while resting.

Many mammals curl up and close their eyes when asleep; birds fluff up their feathers, tuck in their heads, and close their eyes. But animals with a hard exoskeleton and no eyelids can do

none of these things. Playing dead is one defense tactic used by some species against predators that will only eat live, moving prey. If you catch a sulfur butterfly, it will fall into a catatonic state of apparent death and will not react under any circumstance. Once it senses that the danger has passed, it flies off again.

No sex necessary
Many stick-insects reproduce by parthenogenesis. Females do not mate but lay fertile eggs that result in female offspring. In some species, no male has ever been found. The tiny eggs often resemble the seeds of the plants they eat, and when thousands are dropping to the forest floor, they sound like raindrops. In cooler climes, the eggs may lie dormant for a year before hatching.

Seasonal changes

Some insects migrate great distances to escape the rigors of winter. Swarms of monarch butterflies fly south for the winter and their offspring fly back north in the spring. Many insects, larvae or adults, survive winter by entering a state of very deep sleep called diapause. They find a safe place to overwinter, their metabolism slows down, and they neither move nor feed and barely breathe. This same phenomenon exists in tropical climates where the seasons are dry and wet rather than cold and hot.

The nests that wasps build out of chewed wood and

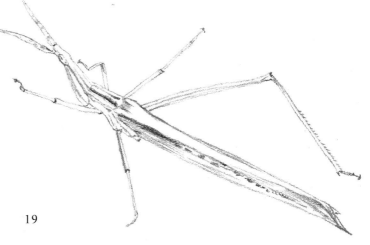

saliva give each wasp a cell of its own (see pages 20–21) during the busy summer season. Each nest, with its perfect honeycomb structure, houses thousands of wasps. What happens at the end of summer? To take the example of a paper wasp (family Vespidae), only the impregnated queen will survive to produce next year's colony. All the sterile workers and now useless males will die. The queen will prepare to hibernate by locating a suitable tree stump where the wood is soft and decomposed. Working with her jaw and legs, she will scoop out a room for herself under the bark, safe from snow and rain, and make it big enough to allow for the debris to expand during freezing. She then lowers her antennae over her face, tucks her wings along the side of her body, and goes into diapause from which she will not emerge until the lengthening days of spring trigger her neuroendocrine system. She may have other neighbors in this old stump: bumblebees, beetles, female earwigs protecting their eggs with their own bodies, and hundreds of ladybird beetles (ladybugs).

Other insects such as the stag beetle larvae, burrow underground and eat all winter long. They spend the winter in a decaying tree, eating dead leaves and rotting wood and getting fatter every day. Some adult butterflies and grasshoppers simply hide in the undergrowth or roll up in dead leaves; if they survive the winter, they will emerge in the spring to breed or lay eggs. Who has not found an apparently dead fly on the window-sill in the wintertime that came to life after the warming sun came through the glass?

Diapause allows the insect to suspend its development until conditions return that are favorable to growth. Hormonal secretions responsible for starting or stopping diapause are created in response to changing environmental conditions. For example, the crop-devastating Colorado beetle spends winter underground in diapause only to become active and emerge when the soil temperature reaches 6.7°C (44°F) in the spring. Other beetle species awaken from winter diapause as soon as the photoperiod (day-length period) increases. Alterations in the photoperiod have a direct effect on the nerve cells in the insect's brain, causing diapause hormones to be released.

Eating the bedding
The caterpillar of the Noctuidae family is nocturnal like the giant moth it will become. The European variety has a wingspan of 10 cm (4 in), while its tropical relative grows to twice that size. The caterpillar remains still and perfectly camouflaged during daylight, using little suckers to attach itself to the underside of a leaf. At nightfall it stirs and stretches before eating the leaf used for its bed.

Resting or lying in ambush?
A praying mantis stands still with head upraised and pincer-like forelegs folded in front, as if praying (see photos left and pages 12–13). It is difficult with such animals to differentiate between a resting or a hunting posture.

Feathered and folded antennae

An insect's antennae are tuned to pick up sounds and smells—vibrations and chemicals that float on the air— so when an insect is actively looking for food or a mate or keeping alert to predators, its antennae are out and searching. When "dozing," one moth will even slide its antennae under its wings. Life is usually short for adult moths and some don't even have mouthparts: their only purpose is to produce the next generation. The saturniid moth above (second picture at top, family Saturniidae) *is one such moth whose vestigial mouthparts preclude any sipping on nectar. The fan-shaped antennae of a scarab beetle (third picture at top; related to cockchafer beetles, also May or June bug) would be closed if the beetle was resting on the ground or just walking around. But when it prepares to fly, it spreads out its feelers to catch the wind direction and any other particles of useful information. Most antlions (left top picture, page 25) are nocturnal and rest during the day with their wings folded over their bodies. In their larval stage, they dig pits under the sand to trap unsuspecting prey. Once the animal slides into the pit, the larva's strong jaws grip it and the insect is sucked dry. They are known as doodlebugs in North America.*

The plants that insects eat can also provide powerful indicators that the season is about to change. Certain leaf-eating beetles are given early warning about the onset of fall when they consume leaves full of carotenoids. They enter into diapause by burying underground, lying still, and drastically reducing their metabolism.

Other environmental factors can trigger shorter periods of dormancy or quiescence as immediate responses to conditions that threaten survival. At the height of a drought or a heat wave, some insects undergo a reduction in metabolism, waiting for the rains to come. These periods of quiescence may be triggered by other factors besides climatic ones. Overcrowding or even cramped habitats may sometimes trigger dormancy. Weevils have been known to enter quiescence if carbon dioxide levels are too high.

Under the scientist's eyes

A group of scientists at the National Institute for Agronomic Research (INRA) located in Orleans, France, has studied dormancy in the larch sawfly. Sawfly numbers depend on the number of larch cones available to eat. If cones are scarce, more

insects remain dormant, waiting for more cones to be available before they become active.

The diapause of the infamous chestnut weevil may vary in length and has been the object of careful study by agronomists anxious to curb this parasite. The female places her egg inside a young chestnut. Once the egg hatches, the larva eats the chestnut from the inside. In the fall, the satiated weevil larva leaves its

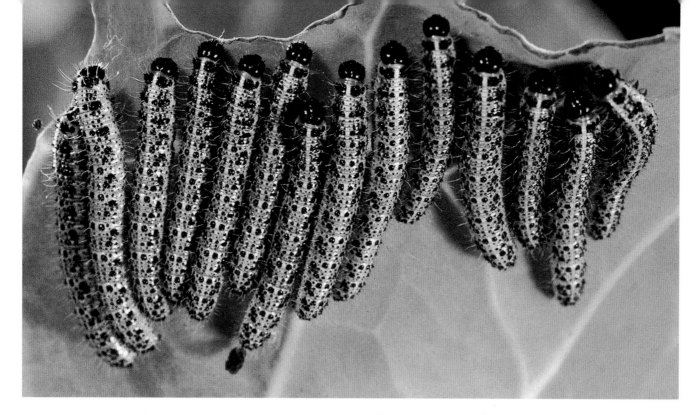

Dining with friends
If a leaf comes under attack from all sides, it cannot defend itself. Usually, when attacked by a leaf-eating insect, the leaf secretes toxic substances from its peripheral pores. Some caterpillars make use of a collective strategy, resting together before setting to work as a group to ensure an abundant meal before the leaf can use its defense system.

Waiting for dinner
The great diving beetle (family Dytiscidae, *photo on near right, page 27) lives in shallow streams and ponds. It rests at night but during the day it is highly predacious, lying in wait for insects, tadpoles, froglets, or small fish. Unlike some water beetles that are scavengers, this is a skilled predator. It must regularly surface to breathe and collect a bubble of air under its wing cases.*

host, buries itself in the ground and begins diapause. The larvae emerge from diapause at the same rate every year to reproduce: 60% after one year, 35% after two years and the remaining 5% a year later. By staggering the end of diapause, the weevil is able to ensure its survival even through unfavorable conditions.

Extended or prolonged diapause has been found in many pests responsible for crop damage such as the pine processionary caterpillar, the Colorado beetle, or the cherry fruit fly. Recent work has revealed that evolution has favored insects that have diapause of varying length, making it easier for a species to survive despite adverse conditions. While life is suspended in this way, the animal slowly consumes its stock of lipids and sugar. When an insect emerges from diapause, just as from hibernation, it will have lost a little weight and will need to get back in shape!

Arthropods in the sleep laboratory

Scientists studying sleep usually carry out their experiments on rats or cats; however, arthropods have also played an important role. Irene Tobler, at the University of Zurich, found that if cockroaches and scorpions were deprived of their daytime rest, they would make up for it the following day by sleeping longer. A German team made similar observations with bees.

The humble fruit fly has done much to further our knowledge of genetics and sleep. By experimenting on this familiar subject in

NO FOOD OR SLEEP FOR THE MAYFLY

Mayflies, distant cousins to dragonflies, are among the most primitive winged insects. Most live for only a few hours once they become adult. At dusk in the summer, swarms of males appear and attract swarms of females. After mating, the males die and the females return to water to lay their eggs before they also die. The larva, however, enjoys a long life of over 20 months (if it doesn't get eaten by a fish), molting sometimes as often as 27 times before finally undergoing metamorphosis and living its short adult life in the mating frenzy.

Last supper

Ticks are arachnids, not insects. This tick (below) can wait for its prey for months, even years (up to 18 years under laboratory conditions). Only a single, precise stimulus will break its inactivity: the smell of butyric acid given off by the sebaceous glands found in mammals. Quickly, the tick attaches itself to its prey, riding along for a few hours before piercing the host's flesh with its rostrum to suck out blood. It is the tick's last supper, however. With its eggs swimming in blood, it drops to the ground to lay its eggs and promptly dies.

the laboratory, geneticists have been able to isolate the gene that governs the biological clock in animals. Research has also revealed that this clock is not solely activated by light passing through the eyes. If a fly is unable to use its eyes, its biological clock is upset, disrupting activity and rest periods. However, little by little, the circadian rhythm settles in once again because the fly is equipped with an emergency "clock" located near the legs and wings.

So, do insects sleep? What we do know is that they undergo different kinds of rest, ranging from what could be called napping or dozing right through to comatose states of the deepest kinds of "sleep" in diapause.

Keeping an eye on time
The multi-faceted eyes of the housefly (above), the horseflies (to the left and second and third photos, top page 28), and the hoverfly (left at top, page 28; family Syrphidae) enable them to not only record nearby movements, but also variations in light intensity. As it notes day switching to night, the insect adjusts its periods of activity and rest. The fly can also rely on the sense organs located on its legs and its small simple eyes (ocelli) on top of its head.

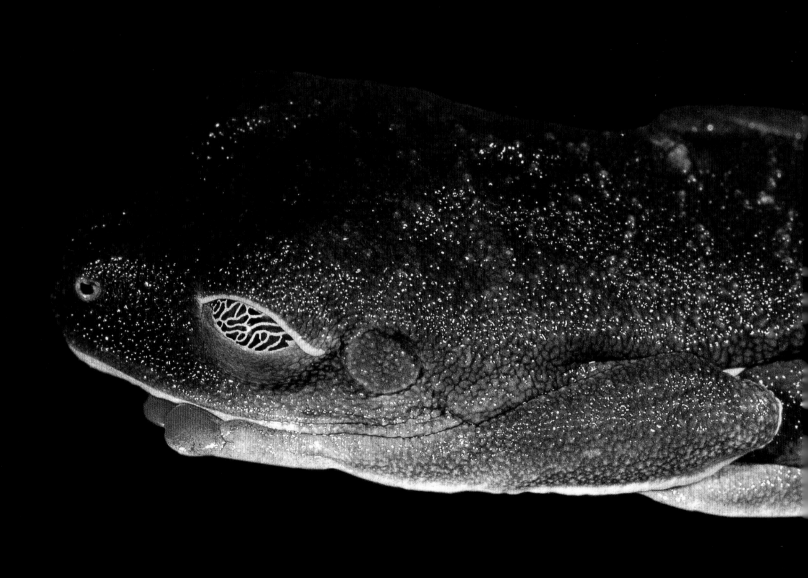

2
Do reptiles dream?

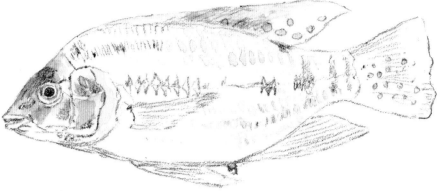

WE HAVE SEEN HOW INSECTS REST and sleep the winter away. But what about other cold-blooded animals such as snakes or lizards? How do you know if a snake is sleeping while its unblinking eyes stare at you? When a chameleon sleeps (see pages 30-31), it closes its turreted eyes and clings to its branch with clamp-like toes, secure that it will not fall. There is a mountain chameleon that lives in the West African rainforest (*Chamaeleo montium*) that marks its special sleeping perch with saliva and returns each night, finding its way back by smell.

The red-eyed treefrog (*Agalychnis callidryas*, pages 32-33) has a nictitating membrane that wipes its eye clean in a blink as it comes up from the lower eyelid (see also pages 48-49). However, when it sleeps, its eyes roll down into the sockets and the bulge of the eye is reduced (see dozing masked treefrog on page 47). Green treefrogs sleep by day; they are bright green when alert and awake but turn a tan color when napping.

We learned how a few insects deal with rest and sleep in the previous chapter, and now we will look at some other cold-blooded animals to see their sleep patterns.

Naps in the water

The green marine turtle resting on the seabed (see left) takes little naps, for it must regularly come up for air. When turtles and tortoises sleep, they actually close their eyes, but they get less than an hour's sleep a day, according to one researcher.

The temperature of ocean waters does not fluctuate much and

Slumbering in the deep
Flatfish (below) blend into sand on the seabed when resting. Both eyes are on the top side in flatfish but while sleeping, they are in an "unseeing" mode. Fish have a sensitive lateral line that picks up any movement around them and this system would probably alert them to danger. Some fish such as tuna have to keep swimming while they sleep or they would drown.

37

stationary (non-migratory) fish stay within a constant range. However, there are pelagic fish that roam the open seas, going from warm waters to frigid ones. Some sharks can actually raise their internal body temperatures considerably higher than the water surrounding them. Contrary to popular myth, many sharks lie perfectly still on the seabed while resting during the day, and some enter such a semi-comatose state that divers can gently touch them without waking them up.

Dr. Colin Shapiro, a Canadian sleep researcher, experimented with guppies (*Tilapia mossambica*) and found that they sleep on the bottom of the tank. When he threw food in, they did not wake up.

The patient snail

Symbolic of slowness, the snail belongs to a primitive group of mollusks already differentiated into some 100,000 species more than 500 million years ago. Only when it rains or after sunset does the snail partially emerge from its shell. A snail's skin is so thin it can "drink" by dipping its foot into water. This fragile skin is no defense against the sun's rays and to prevent drying out, it stays indoors during the day and forages at night. If a heat spell persists, the snail closes the opening to its shell with a film of mucus (slime secreted from its mouth) and waits for dew or rainfall. The desert snail breaks all records for estivation. It can survive without food or water, remaining dormant in its shell for up to five years. In winter most snails hibernate by staying inside their shells.

But does it really sleep, curled up within its coiled house? Researchers at the Institute for Agronomic Research in Rennes, France, have perfected breeding techniques for *Helix aspersa*, a snail native to North Africa. Their research has shown that although hibernation is not a necessity for survival, a daily cycle of night and day is essential (active at night and at rest during the day). Snails live at a slow pace so it is not always easy to tell when it is "sleeping"; however, if its foot is left trailing outside and there is movement within its shell, it may be merely thinking about its next step.

The third eye

Chronobiologists are specialists in the study of natural rhythms of rest and activity. They are particularly interested in the "third eye" (the pineal gland) found in fish, amphibians and reptiles. This gland secretes melatonin and is the internal clock that dictates periods of rest and activity to other organs. Diurnal animals receive stimulation at the end of the night, thus preparing the organism for waking and the day's activities. Conversely, at dusk the internal clock signals a slowdown in metabolism, and cardiac and respiratory rhythms diminish, readying the body for the onset of sleep. Of course, in nocturnal animals this process is reversed. Many nocturnal geckos (sketch above) spend their daytime hours under bark or stone, absorbing heat by conduction. The length of their basking period is controlled hormonally by the pineal gland. In many diurnal lizards this third eye is not covered by bone but lies under an opening in the skull, covered by a translucent scale. Highly developed in fish, and sometimes even visible through the skin at the top of the skull, this gland is easy to remove for study

At the Ethology and Neurobiology Laboratory at the Louis-Pasteur University in Strasbourg, Michel Anthouard chose to study biological rhythms of the common goldfish, an archetypal fish. In the 1970s, he perfected a self-feeding technique whereby the fish obtained food by prodding a metal rod. The number of impacts on the rod records the fish's activity. Anthouard also worked with bass documenting a correlation between an increase in water temperature and activity in fish.

Estivation versus hibernation
Snails estivate in the summer by slowing down metabolic functions, though not as drastically as during hibernation. In winter the snail shuts down almost completely. Protected by a thick, opaque, tightly fitting curtain, a snail breathes one hundred times slower than during periods of activity.

39

A fish's bed
Fish prefer to hide in the mud or behind seaweed, coral or anemones to sleep. Some lie low between pebbles or other crevices, and others, such as the moray eel, have deep burrows where they rest.

If the water temperature drops below 20°C (68°F), the giant catfish, for example, eats less and stops eating completely if the water temperature drops below 12°C (54°F). Each species of fish becomes lethargic at a given temperature. Anthouard noted that fish will eat greedily to make up for fasting.

Although fish living in polar seas are protected by fatty tissue, herring and smelt produce anti-freeze proteins. Fish such as carp, hide away in silt on the river bed and reduce their metabolism. A few tropical species, such as the lungfish, estivate during the dry season by burying themselves in the mud.

Toad goes courting

In northern climates when the days begin to shorten, toads, like most amphibians, go into a state of dormancy often called hibernation. Some toads dig under a heap of leaves or burrow

SLEEPING BAG

Parrotfish prepare for sleep each evening by finding a narrow crevice among the corals and secreting a bubble of mucus around themselves. This transparent covering solidifies in water, making a sleeping bag. A diver can gently pick up such a cocooned parrotfish without waking it. Parrotfish often have favorite resting places for sleeping.

underground. While in their excavated shelters, their breathing, digestion, circulation and excretion all slow down. The same process happens in hot climates where the toad estivates by burying itself in the mud and waiting for the dry season to end.

As spring approaches, the male toad is transformed. His testes start brimming with semen, and his legs develop nuptial pads that will enable him to grasp the female during amplexus (the position for mating; see page 46). On emerging from hibernation, the mating instinct takes hold immediately. The toad makes its way to a pond or ditch to call for its mate. Water is essential for amplexus and for the subsequent development of the eggs.

Except during the mating period, adult toads and frogs are most active at night while the young are far more active during the day. While females are more likely to roam, males rarely leave their territories. In the face of danger, toads often remain motionless no matter how uncomfortable the position or how long they have to hold it. Despite the fact that amphibians may spend a long time in dormancy, some scientists remain dubious as to whether they really sleep. However, one research report stated that the western toad sleeps for 14.5 hours a day.

A sleeping snake

Turtles are truly ancient—the earliest fossil dates from the Permian period, about 280 million years ago—but snakes only appear in the fossil record about 140 million years ago, which makes them relative newcomers. But what a success story! Today there are about 2,600 species of snake in comparison to only 260 species of turtles and a mere 20 species of crocodile. The snake has adapted to a wide range of climates, from the equator up to the Arctic Circle.

It is difficult to say whether a motionless snake is awake, resting, lying in ambush, digesting or basking. Snakes have no eyelids and can therefore never close their eyes. But they can rest motionless for hours—and they do yawn.

Unlike other reptiles, snakes are not restricted to a diurnal-nocturnal rhythm. Their habits may vary depending on the seasons. The desert horned viper from the Sahara (see page 51) normally is active during the day but becomes nocturnal in

Safety in numbers
Pelagic fish swim in the open sea and do not take cover. Grouping together in a shoal provides natural protection. When resting, the shoal remains stationary. A few discrete fin movements enable each individual fish to stabilize its position (see pages 42–43).

Using a swim bladder to sleep
Bony fish have a swim bladder that is used to achieve neutral buoyancy, neither floating to the top nor sinking to the bottom. To inflate its swim bladder, a fish swallows enough air at the surface to reach the desired depth. It only has to flush out the air in bubbles to go deeper. This system allows fish to sleep without swimming. Water is still being pumped through the gills for oxygenation. Sharks, which are cartilaginous fish, have a large fatty liver that serves a similar purpose.

FROGS WITH ANTI-FREEZE

Some amphibians can survive in freezing temperatures, particularly the wood frog (*Rana sylvatica*). How is this achieved? Bill Schmidt from the University of Minnesota solved the mystery about 20 years ago. He was amazed to discover wood frogs frozen stiff. When temperatures drop to below freezing, *Rana sylvatica* begins to synthesize glucose, a natural anti-freeze. Glucose prevents ice crystals from forming, which usually causes death due to tissue rupture. Intercellular tissues are slowly held in ice, while the internal organs are saved from damage. After a freeze of two weeks, frogs have been successfully thawed out. Advantages to freezing are that frogs can emerge from hibernation as the snow is melting, thereby breeding early in these temporary ponds; a sudden cold snap will not kill them; they can expand their range further north.

summer to maintain its temperature at a constant 28°C (83°F). Buried under the burning sand during the day with only its head breaking the surface, the snake remains on the alert.

Like other ectotherms, snakes use the warmth of the sun to raise their body temperatures. After the asp leaves its nocturnal burrow, it basks coiled up on a tree stump in direct sunlight. An hour in the morning sun is sufficient to reach 28°C (83°F), the minimum temperature for activity. Once the snake has warmed up it can catch an amphibian or small mammal and digest its prey. But without heat, it cannot digest. If the sun disappears behind heavy clouds for too long, the viper will regurgitate its prey to prevent undigested food from decaying in its digestive system. If the temperature drops below 10°C (50°F), the viper is unable to feed and has no alternative but torpor or dormancy.

When the hours of daylight decreases sufficiently in temperate climates, nearly all snakes will hibernate for two or three months. In boreal or mountainous zones, some snakes will hibernate up to eight months, intertwined with their neighbors in huge dens below the frost line.

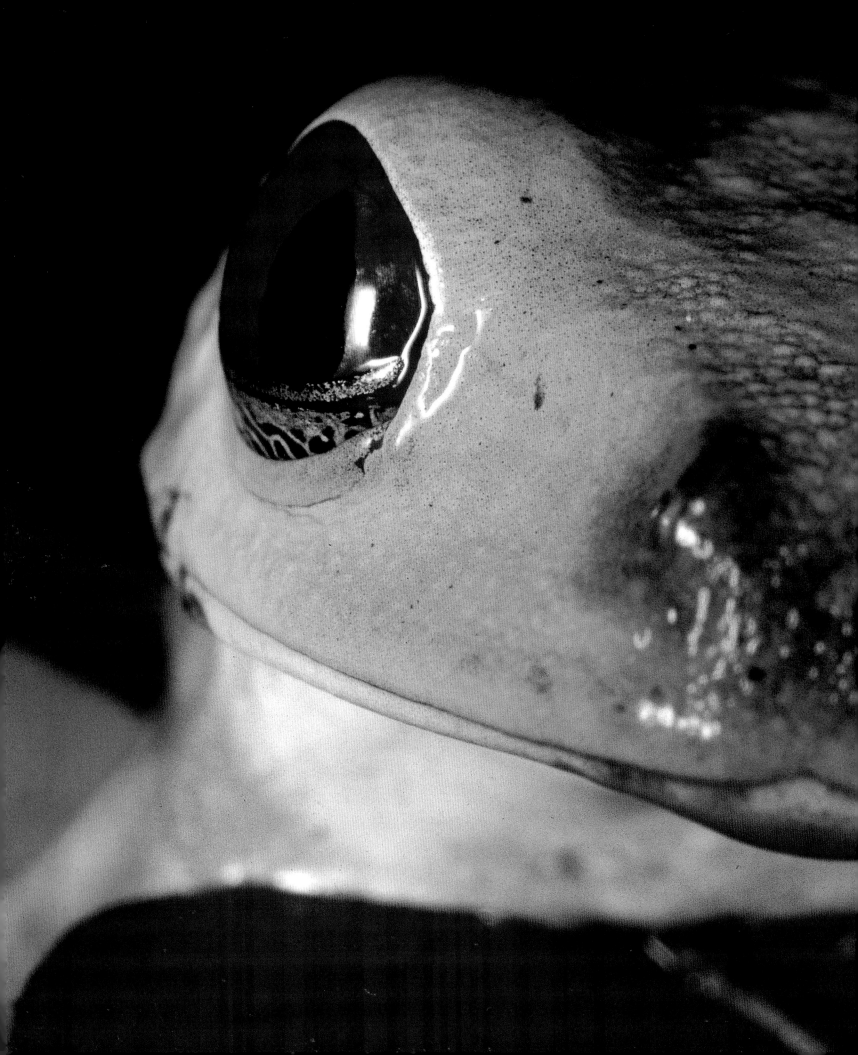

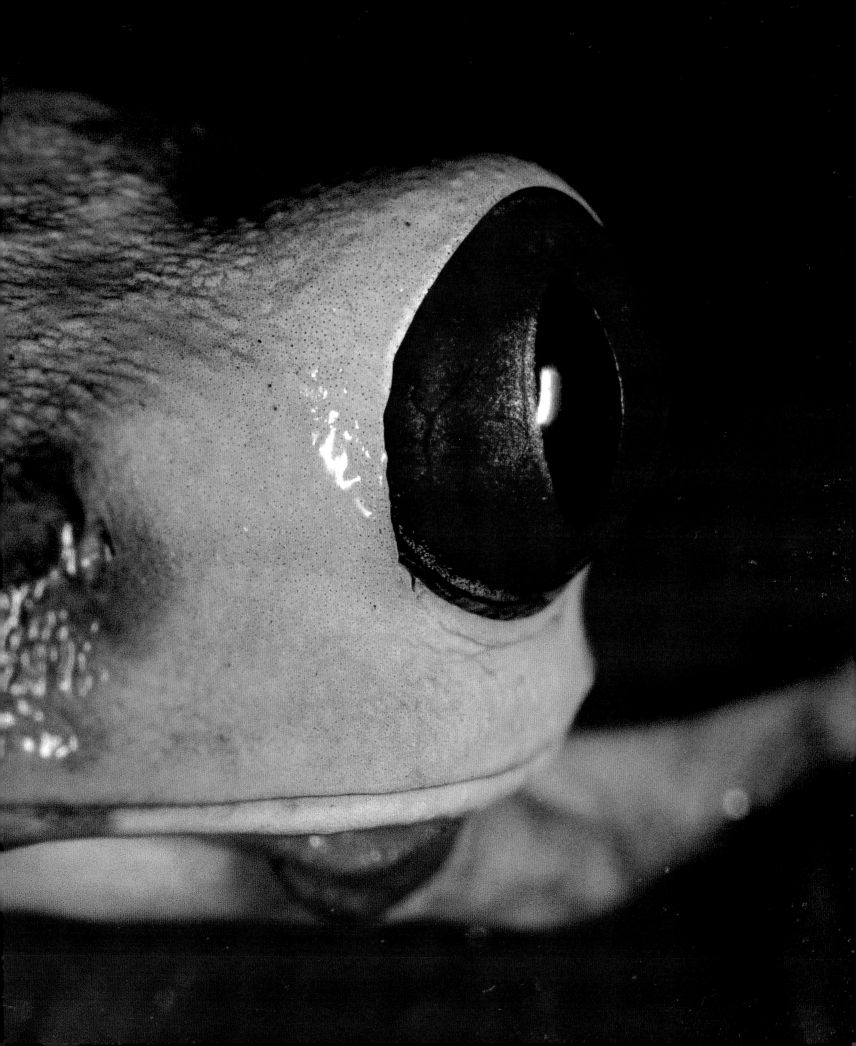

Playing dead
In the face of danger, grass snakes will feign death. Remaining quite still with mouth open and tongue drooping (above), they will even emit a fetid odor from their anal glands in an unbridled attempt to simulate death and avoid being eaten. The hog-nosed snake can even be picked up and turned over but it will immediately flop onto its back and continue to play dead until you disappear.

If one turtle yawns, is it catching?

Turtles are very popular as pets but people also want to protect endangered species such as the leatherback turtle or Hermann's tortoise. Bernard Devaux has founded a sanctuary in Gonfaron, France—Station d'observation de protection des tortues des Maures (SOPTOM)—to protect and reintroduce the native Hermann's tortoise (*Testudo hermanni*). Turtles are peaceful creatures, solitary and independent. They don't like barriers and will go to any lengths to escape captivity. However, if they are left unhindered and have all their needs filled, they rest contentedly.

50

Will it fit?
This common garter snake is eating a fish that is several times the size of its head. A snake's top jaw is loosely attached by ligaments to the cranium so it can stretch downward, and the bottom jaw is not fused in the middle so can expand sideways. After it eats, it must rest for several days or weeks (for large snakes) to digest. As snakes do not have eyelids, it is hard to say if they sleep while digesting but different research reports state that pythons sleep from 15 to 18 hours a day.

Like most tortoises, the Hermann's tortoise is diurnal and must spend hours basking in the sun. If the temperature goes below 10–13°C (50–55°F), Hermann's tortoise would have to enter a dormant state. At sunset, it slowly returns to its nighttime shelter, where it remains motionless, limbs tucked in, head sunk down, eyes closed. Each morning upon waking, the turtle yawns, stretches its neck and legs, and sighs deeply. Some observers claim tortoises dream. Certainly they have unexplained jerks during their sleep. Devaux has discovered that their eyes can sometimes be seen moving wildly beneath eyelids. But can it be classed as REM sleep? The subject remains open for debate.

Experiments in freezing

In cold climates some amphibians and reptiles overwinter in the mud of a pond or in underground burrows to avoid freezing temperatures. Turtles that go into the water stop breathing with their lungs and instead do minimal breathing through skin and cloaca. Some turtles and amphibians hibernate near the surface, under a pile of leaves exposing themselves to more extreme freezing temperatures. Encephalograms have recorded brain activity in frogs and the cessation of this activity when a frog is frozen.

Dr. Kenneth Storey, professor of biology and molecular physiology at Carleton University in Ottawa, heads up a team that studies freeze-tolerant vertebrates such as the wood frog (see box page 46), the garter snake, painted turtles and box turtles—all of which have the ability to withstand the freezing of

Hiding in the sand
Snakes depend on their coloration for camouflage, not only to surprise prey but also to hide from predators. The desert horned viper (above) blends into the desert sand. It is also a bit cooler under the sand than on top of it so this burrowing can be used to regulate its temperature.

GROUP HIBERNATION FOR GARTER SNAKES

Snakes often hibernate in groups with different species all using the same underground den or hibernaculum, year after year. Ideal underground chambers have temperatures between 40°F (4°C) and 52°F (11°C). Breathing and metabolism slow down, and not much weight is lost during hibernation. The record for mass hibernation is held by a Canadian site: 10,000 garter snakes in the same den! Hibernating together in this way limits heat loss and provides protection against dehydration. It is also convenient to find mates quickly when emerging from hibernation in the spring.

as much as 65% of their body water and still be able to thaw out unharmed.

In freezing temperatures frogs assume a crouched position with limbs tucked in, similar to the "water-holding" stance of frogs undergoing heat stress. There is no heart beat, no blood circulation, no breathing and no detectable brain activity. However, after thawing out, all vital functions return unimpaired within one or two hours after a deep freeze that may have lasted up to two weeks. Contrary to hibernating mammals, amphibians and reptiles lose little weight during their winter lethargy.

Other animals Dr. Storey has examined are the gray treefrog that turns blue when frozen; painted turtle hatchlings that overwinter in their nest and tolerate being frozen; garter snakes that can be frozen for up to a day; and box turtles that can survive two days frozen at –2°C (28.4°F).

A crocodile's smile

A crocodile spends hours motionless with its eyes at water level. Despite appearing like a floating log, it is not sleeping. Should a careless water bird or an antelope approach to quench

its thirst, the crocodile will attack at breathtaking speed. Research on crocodiles in the lab is scarce because they do not adapt well to captivity and their habits become modified, particularly their sleeping patterns. However, since they are related to birds—and REM sleep occurs in birds—sleep specialists have taken a keen interest in them. Some scientists believe that sleep appeared first in reptiles even if dreaming is absent.

Like insects and other ectothermic animals such as mollusks, fish and amphibians, reptiles rest and undergo varying periods of inactivity depending on their circadian rhythm and climatic conditions. Certainly they sleep, if for much shorter periods than do mammals. The capacity of some species to undergo freezing revolutionizes our concept of hibernation, and perhaps someday we will discover whether crocodiles dream.

Yawning or cooling?
Crocodiles often rest with gaping mouths to maintain their internal temperature. If turned toward the sun, they are allowing the warmth to reach areas where blood is near the surface of the skin. Open mouths also work as a cooling method, allowing heat to dissipate. If the outside temperature rises too sharply, however, both the American alligator and the Nile crocodile enter periods of extreme lethargy, with a sharp reduction in metabolism. If an alligator experiences freezing temperatures, it rests in shallow waters with its nose above the surface for breathing. Under laboratory conditions, one researcher found that crocodiles sleep about three hours a day, with their eyes closed.

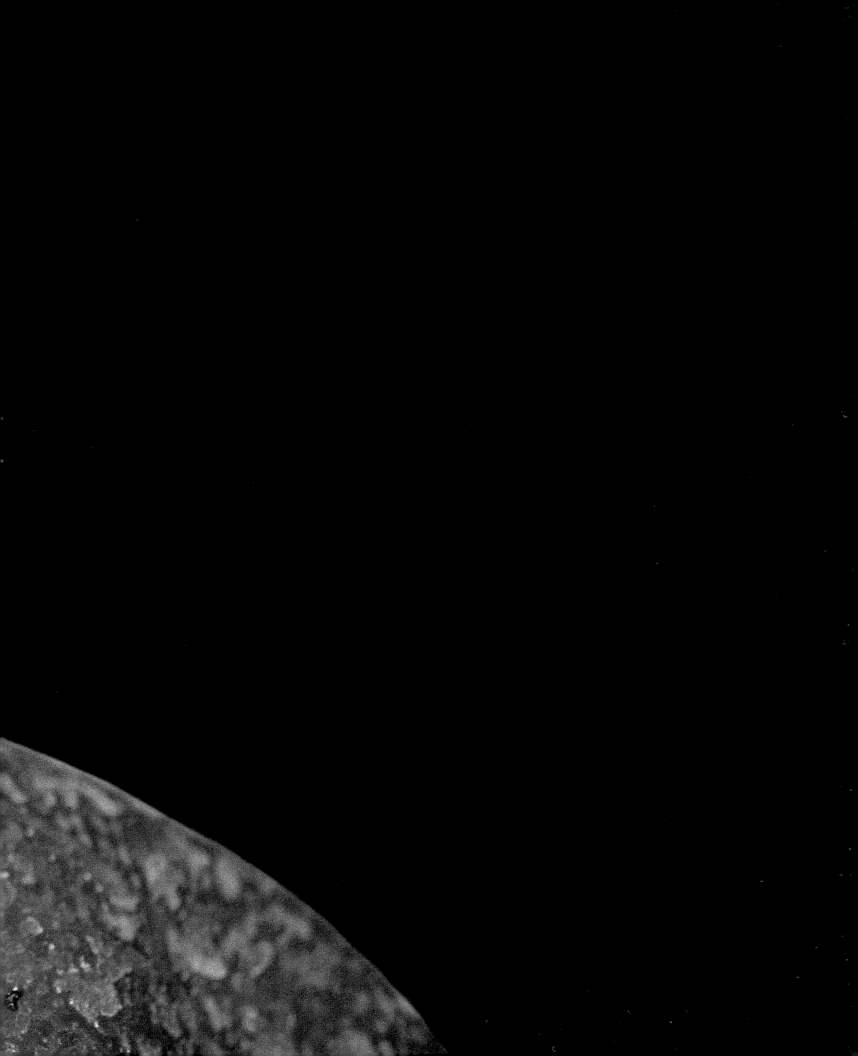

3

Do birds sing in their sleep?

SLEEPING POSTURES OF BIRDS VARY from crows roosting in the treetops to nighthawks on the ground to the flamingo standing on one leg, neck corkscrewed backward, beak tucked under wing. Whether they are diurnal or nocturnal, sleep upright, crouch down on land, or bob on the water, all birds examined so far exhibit deep slow wave sleep, followed by a short burst of rapid eye movement (REM) sleep or paradoxical sleep, which researchers believe indicates dreaming in birds as it does in mammals. Birds such as owls, which cannot move their eyes, emit the typical brain wave pattern of paradoxical sleep without the actual eye movement.

In reptiles, amphibians and fish, we saw how the "third eye" (pineal gland) reacted to light and directed the circadian rhythm of these animals. In birds, the pineal gland is situated in the brain, and light coming in through the eye triggers this gland to produce a hormone that dictates when birds go to sleep and when they wake up. As the amount of daylight increases, birds sleep less. Stanley Coren, a Canadian psychologist who has written on sleep research, says that on average a diurnal bird sleeps around seven hours in a 24-hour period. However, the sleep is taken in short naps, sometimes only a minute long for birds that sleep on the wing, and the paradoxical sleep of birds is rarely longer than 30 seconds and usually only 5 to 6 seconds long.

On the evolutionary trail

Birds evolved from reptiles about 150 million years ago, whereas the fossil record shows the first mammals evolved from therapsids (mammal-like reptiles) about 220 million years ago. It is interesting to speculate on how the mechanism of sleep evolved; researchers may someday provide the answer. No REM-type sleep waves occur in reptiles or amphibians, so what, then, triggered the bird and mammal brains to develop this pattern?

Patterns of sleep
Although there are about 9,600 species of birds, certain sleep behaviors are common to most. Many birds roost in trees, some singly, some as a colony. Nocturnal birds such as the barn owl (pages 54–55) hunt at night and sleep during the day, perched on their roost, eyes closed, feathers fluffed out for warmth. Budgies (pages 56–57) are active mainly in early morning and late afternoon and at night roost in trees, eyes closed, beak buried in warm feathers. Hummingbirds (pages 58–59), high-energy creatures, enter a nightly torpor to conserve energy. They don't so much sleep as enter a shallow torpor. The flamingo (photo on left) stands on one leg, head twisted back and beak under its wing.

Guarding the flock

Charles Amlaner and his team from Indiana State University showed that some birds, including the mallard (photo below), have unihemispheric slow wave sleep. Under conditions where they feel safe from harm, both hemispheres and both eyes close. If they feel vulnerable, they sleep one hemisphere and one eye at a time. When birds sleep as a flock, birds on the fringe act as lookouts, sleeping with one eye trained on the horizon. These "watchbirds" alert the group instantly and can take off in a split second. The scarlet macaw (page 63) sleeps with its flock, roosting in trees or nesting in holes in tree trunks. This one is half-alert, with one eye on the lookout.

Equally fascinating is the origin of the unihemispheric sleep that birds use when they need to be alert to danger.

Charles Amlaner's team of researchers at Indiana State University used electroencephalogram (EEG) recordings to confirm how birds achieve unihemispheric sleep with their two hemisphere brain: when one hemisphere is sleeping and showing slow wave sleep, the eye controlled by that hemisphere (opposite side) is closed, but the other hemisphere is awake and its corresponding eye is open and alert.

When ducks sleep in a row, the ones on the edge keep one eye open—the eye facing the outside where predators lurk—and immediately fly away if they spot an enemy. They also rotate their bodies so that they can watch with the other eye and each hemisphere get some sleep. Research also showed that the ducks on the edge spent 2.5 times longer in unihemispheric sleep than did the ducks in the center.

Researchers at Amlaner's lab compare their findings of unihemispheric sleep in birds with that of some lizards who slept with one eye open after seeing a predator. They postulate that perhaps all animals were once able to sleep one hemisphere at a time but those that acquired safe places to sleep lost the ability

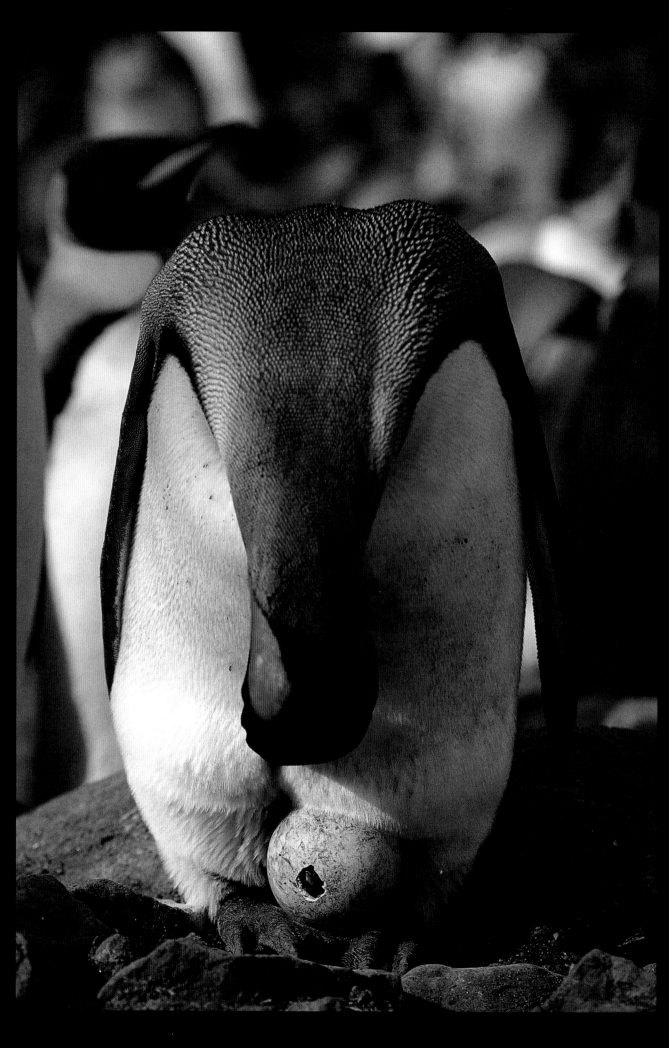

ASLEEP ON ITS FEET!

The king penguin incubates its egg by keeping it on its feet and covering it with a warm abdominal fold of flesh and feathers. It lives in colonies on rocky islands, but the emperor penguin lives on ice packs so if his egg ever rolled off his feet, the egg would freeze in minutes. With emperor penguins, only the male incubates the egg while the female goes back to the sea until the chick hatches. However, with king penguins, both parents incubate the egg. Each parent takes a three-week shift and does not eat, drink or move during that time. They sleep in short naps of a few minutes, standing upright, head hunched, flippers by their sides. The chicks cluster in crèches while parents hunt for food that is regurgitated into the chick's throat. The king penguin chick overwinters with its parents as there is not enough time to fledge during the short summer.

because it was no longer needed. We will see in the next chapter how certain aquatic mammals have retained the ability to use unihemispheric sleep.

Researcher Neils Rattenborg from the Amlaner lab further notes that under severe stress, humans exhibit sleeping brain-wave patterns similar to the unihemispheric sleep of birds. Did the trauma "wake up" an old part of the mammalian brain geared to staying alert when danger threatened?

Birdwatching

The electroencephalogram has enabled sophisticated research to be carried out in the laboratory. However, natural conditions for the animals must still be maintained, and it is essential to leave sleep patterns as undisturbed as possible. Chickens and pigeons are often used for experiments, and they sometimes undergo small operations for placement of electrodes, thus enabling EEG scans to be made. In this way, we can measure variations in cerebral activity between wakefulness and sleep. In the field, daily activity rhythms or migrations can be recorded by fitting birds with transmitters. How can the activity of a small bird flying high in the sky be recorded in detail? Has it just closed its eyes? Is it blinking or sleeping?

Birds have developed a specific type of sleep, interrupted by short bursts of alertness. When a bird falls asleep, its reaction threshold to stimuli is significantly diminished, muscle tone is reduced, breathing slows down, and cardiac rhythm decreases. Changes in EEG waves are proof of the depth of sleep. As early as 1963, Marcel Klein, a French biologist on Michel Jouvet's team, collected data revealing the existence of REM sleep in pigeons and chickens.

Up until now, all the birds singled out for observation by researchers—vultures, Mexican eagles, ravens, owls, king penguins, to name a few—have shown a marked difference between wakefulness (rapid waves) and sleep (slow waves) when EEG waves have been compared.

All had short bursts of REM sleep (10 to 20 seconds) following their slow wave sleep. Similar to young mammals, chicks show about five times more REM sleep than their parents. However,

Back to earth
Birds such as swifts and albatrosses sleep on the wing. A royal albatross was once tracked for almost two hundred days in the air, without ever once coming to land. This white pelican (above) sleeps on the ground in the posture we consider typical for birds: eyes closed and beak tucked under its wing.

Fasting and sleeping
When food is scarce, many birds fast and sleep for longer periods of time. Both fasting and sleeping are energy-saving tactics. Under laboratory conditions, birds forced to go without food slept longer. In 1984 Gerard Dewasmes, a French research scientist, showed a correlation between tolerance to fasting and sleep duration. Both adults and young of the common swift will enter torpor, with a slowed-down metabolism and a lower temperature. Hatchlings can endure up to 10 days fasting with a 50% weight loss—and a subsequently longer time to fledge.

avian sleep is characterized by a small proportion of REM sleep, only 3 to 6% of total sleep, compared to 25% for the bird's arch enemy, the cat. EEGs carried out on geese revealed many snatches of REM sleep, with up to 200 recorded in a 24-hour period.

In 1968, Russian scientist Ida G. Karmanova, from the Sechenov Institute (Saint Petersburg), noted this particularity in avian sleep. Karmanova defined four separate states in hens: wakefulness (33% of the time), slow wave sleep (44%), REM sleep (5%) and day rest (18%). Day rest is known as cataleptic rest since the bird appears to be paralyzed, while still being aware of its surroundings. Karmanova also described a sleep particularly specific to nocturnal birds such as the owl, interrupted by short periods of wakefulness with eyes open at the end of REM sleep. These short bursts of alertness (5 to 7 seconds) enabled the bird to remain alert to any lurking danger.

We have seen how birds can control unihemispheric sleep and how they can alternate between periods of sleep with both eyes closed and periods during which they are "half-asleep" with one eye open. During these periods of semi-rest, the EEG waves in the alert hemisphere are similar to that of wakefulness, allowing the bird to remain alert while taking some sort of rest. When conditions are safer, this unihemispheric sleep gives way to continuous sleep with eyes closed. In the laboratory, free from the fear of predators, open-eyed sleep rarely occurs.

Sleeping on the wing

The common swift is an aerial phenomenon. Not counting the time spent laying and incubating eggs and rearing the chicks, the swift spends 90% of its life on the wing; it even mates in the air. While in flight it gathers the necessary material for nest building by catching in its beak whatever twigs, pieces of string or plant debris the wind tosses up. To feed, the swift opens its huge mouth wide and devours all the insects that cross its flight path. When food is scarce, swifts can enter a state of torpor in order to survive. Both adults and young can last up to 10 days without food without ill effects.

Every year the common swift migrates about 500,000 km (312,500 mi) in the eight months between the time it fledged to

Dreaming of other times
Scientists theorize that birds evolved from a small carnivorous reptile that lived during the age of dinosaurs, which ended 65 million years ago. Birds do not sleep a lot but their brain waves show both slow wave sleep and REM sleep.

Asleep on the snow
The sleeping crane (see page 68) stands on one leg, head immersed in plumage. Mute swans (pages 66–67) are as relaxed sleeping on snow as if it were summer grass. Feathers provide excellent insulation against the cold, and while one guardian stays alert, the others sleep and dream.

Sleeping while the bells toll

Barn owls (Tyto alba) are graceful birds of prey that are found worldwide except for parts of temperate Asia and many Pacific islands. Their heart-shaped face has earned them other names such as monkey-faced owl. They live alone or in pairs and roost during the day in farm buildings, hollow trees or caves. Barn owls not only like to nest in barns, they also inhabit church belfries and can be found during the day in a church tower. Apparently, the sound of church bells tolling doesn't disturb them—they don't bat an eyelid. Barn owls favor the same nesting site, year after year. Hunting at night for rodents and small birds (see page 71), a parent brings supper for its chick.

its return the following summer to its nesting site. It is not a perching bird, for its four toes all point forward, but it can cling to the sides of plants or trees. A discovery by Swedish researchers in 2000 showed some swifts roost like bats, hanging from willow branches by their claws.

Whole generations of ornithologists thought the swift never slept but now we know it sleeps on the wing. At night swifts settle at an altitude of between 915 and 1,828 meters (3,000–6,000 ft), above a pocket of warm air, and begin a rest flight, flapping their wings every 4 seconds then gliding for the next 3 seconds.

It would seem that penguins sleep in a similar fashion when they migrate for long periods, swimming as they sleep. They sleep

in regular bursts, waking to breathe at the surface from time to time. On land they sleep standing up.

Sleeping on a perch

While some birds sleep head upright, like the owl, or with head retracted, most tuck their heads or at least their beaks under a wing to breathe air warmed by their plumage. Standing on one leg is probably an energy-saving device for many birds as the folded leg rests in the warmth of the abdomen feathers.

Why don't sleeping birds fall over? Their center of gravity is located very low, just above the feet. When perching birds settle down for the night on a branch, they lower their body, bending their legs. As they do this, tension in the tendons triggers a locking mechanism in the toes. In addition, a knob on the upper end of the femur (thigh bone) presses against the ileum in the pelvis, locking the leg in place. The sleeping bird rocks horizontally ever so slightly to keep its balance.

In 1984, Gerard Dewasmes reported how the neck muscles in the goose relaxed during REM sleep. Until then, ornithologists had observed only a reduction in tonus and had concluded that birds do not show muscle relaxation as is the case in mammals. In fact, it all depends on position during sleep. Chickens and pigeons, the main birds used in laboratory research, sleep with their heads under their wings. Lost muscle tone during sleep causes them to relax the wing that holds their head in place. The bird must therefore contract its neck muscles from time to time to keep its head in position; hence, the variable tonus recorded during sleep. When Dewasmes studied the domesticated goose, which sleeps with its head on its back, he noted a drop in tonus of the neck. Only birds such as the gull or flamingo, which sleep with heads supported, can completely relax their necks. In other species, muscle relaxation is staggered.

Birds that are born with eyes open and feathered are called precocial or nidifugous; they don't sleep as much as do the altricial or nidicolous chicks that are born naked, eyes closed,

One for all and all for one
*Little bee-eaters (*Merops pusillus argutus, *pages 72–73) have a complex social structure, forming clans within roosting flocks, with youngsters from one brood staying on as helpers with the new chicks. They have an almost ritualistic sleeping behavior: they arrive at the roost before dusk and shuffle into line on a perch, pressing their sides together. When everybody is settled in, they close their eyes and go to sleep. They also clump together like this during the day if the weather turns cold.*

Home sweet home
*Most birds build nests to raise their chicks,
not as roosts. The eggs need to be incubated
and the chicks fed in a safe enclosure. Once
chicks are fledged, the nest is not usually used
and birds go back to their normal roosting
perches (see exception on page 76). Some
birds use hollow trees (below) and others
seize the opportunity of a man-made structure
with built-in heat (above). The male golden
palm weaver (Ploceus bojeri, page 75) is an
accomplished architect that weaves an intricate
structure using his beak and feet.*

totally helpless and dependent on parents. Altricial hatchlings spend their early life sleeping and eating what their parents bring, but precocial chicks are out of the nest walking and swimming, finding their own food within a day of hatching.

Torpor and hibernation

When cold weather hits, small birds are vulnerable to freezing at night when they go to sleep. They must try to get enough food during the day to have some fat reserves for the night. An additional survival tactic that a number of birds use is to go into a torpor that can last the night or for a few days. Research on titmice found that in a state of torpor their temperature dropped to about 29°C (84°F) but they cannot do this for more than a couple of days or they will die. By overlapping their feathers and pressing close together, long-tailed tits conserve up to 38% of their body heat. Hummingbirds go into nightly torpor for about 12 hours to conserve energy and their temperature drops from 41°C (106°F) to a low of 10°C (65°F). Upon waking a humming-bird must shiver to warm up its tiny body. It may take up to an hour of shivering before its temperature is warm enough to allow it to seek out breakfast.

Ralph Berger has shown that at the onset of torpor, a stage of slow wave sleep occurs. During torpor, birds will sleep significantly longer but will essentially remain semi-alert. Only one bird known to scientists truly hibernates: the common poorwill (*Phalaenoptilus nuttallii*). This small, short-tailed nightjar flies around at night hunting moths and beetles. By day it roosts in shrubbery and at dawn or dusk, it sounds its repetitive "poor-will" song. In October it seeks out a rock crevice, to which it returns yearly, and wedges itself in for winter hibernation. Its heart rate and breathing drop to almost undetectable levels and body temperature falls from 40–41°C (104–106°F) to 18–19°C (64–66°F). When daylight hours announce the arrival of spring, the bird awakens.

"Half-asleep" or "half-awake"?

Ornithologists have described several species as having short naps with one eye open. Signs of sleep have been recorded in the

74

Birds of a feather nest together

*Sociable weavers (*Philetairus socius*) of Africa build huge communal nests of straw and plant stems that can measure over 3.6 meters (12 ft) across and 6 meters (20 ft) high with as many as 150 bedrooms! They live and roost in these nests and do not use them just for breeding. Each monogamous pair has its own entrance and bed chamber. All the birds build the common dome-shaped roof together before constructing their individual "condo." The same nest will be used for years, with regular repairs and new "couples" adding on their own chamber. As the whole nest gets bigger every year, it sometimes becomes so heavy that it crashes to the ground and they have to begin anew.*

hemisphere of the brain that corresponds to the closed eye (i.e., left hemisphere and right eye) and signs of being awake in the other. Scientists have concluded that birds sleep with one eye closed when they are vulnerable.

In fact, researchers disagree on the definition of avian sleep. Charles Amlaner is convinced that birds can sleep in "unihemispheric sleep" like dolphins. Berger talks more about a "half-awake" state. Gerard Dewasmes disagrees with this "half-asleep" concept and states: " 'Half-awake' birds are most probably in a

sleeping state but it would be wrong to talk about a half-sleep or even less, real sleep. Otherwise birds would spend 50% of their time asleep! Let's restrict ourselves to sleep with eyes shut. Under optimal conditions in the laboratory, sleep rarely exceeds 25 to 30% in this case."

Birds present other singular features. Birds of prey such as the eagle or falcon have a level of REM sleep identical to starlings which are their prey. In mammals the level of REM sleep differs depending on whether the animal is predator or prey. In contrast to mammals, large birds such as the penguin show more REM sleep than small ones. The small amount of sleep needed by birds might be because they do not lie down to sleep like mammals. If birds had completely flat muscle tone, they would fall out of trees. According to Israeli scientist Peretz Lavie, these particularities suggest that sleep type has developed as a response to the particular conditions of a given species.

Hunger and torpor
When the common swift have to leave their young for a few days without food, the chicks' metabolism slows down, their temperature drops to as low as 4.4°C (40°F), and they enter a state of torpor for up to 10 days. When their parents return with food, the young revive and warm up in less than 30 minutes (see below).

NIGHT OWL OR MORNING LARK?

Early in the morning in southern China, the Peking robin (*Leiothrix lutea*, top right) sings a song so melodious that it is often captured, caged, and given as a wedding gift. Not so the nocturnal oilbird (*Steatornis caripensis*, above), which roosts in dark caves and has a call more akin to a harsh scream or squawk. The oilbird uses echolocation to fly about its cave, and its large eyes are specialized for night vision. A fruit eater, it swallows fruit whole and when it comes back to its roost, it dozes and digests the whole day through. New Zealand's kiwi bird (page 79) is also nocturnal, and during the day it sleeps in a burrow, often under tree roots.

Singing in their sleep

Biologist Daniel Margoliash at the University of Chicago has shown that the brain cells of zebra finches show the same kind of electrical activity when they sleep as when they sing. He tracked individual neurons while the finches were both awake and sleeping. He saw that they fired in the same patterns when the bird was awake and singing as when it was sleeping while researchers played a recording of its song. These same patterns do not appear when the bird was awake and listening to a recording of its song. When the bird slept in silence, the neurons still fired in patterns similar to those when it sang but somewhat distorted. Margoliash speculated that the birds were rehearsing their tune, strengthening the pattern in their "dreams."

4
Do all mammals dream?

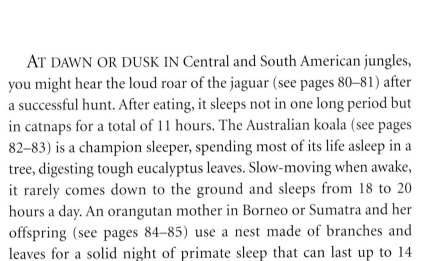

AT DAWN OR DUSK IN Central and South American jungles, you might hear the loud roar of the jaguar (see pages 80–81) after a successful hunt. After eating, it sleeps not in one long period but in catnaps for a total of 11 hours. The Australian koala (see pages 82–83) is a champion sleeper, spending most of its life asleep in a tree, digesting tough eucalyptus leaves. Slow-moving when awake, it rarely comes down to the ground and sleeps from 18 to 20 hours a day. An orangutan mother in Borneo or Sumatra and her offspring (see pages 84–85) use a nest made of branches and leaves for a solid night of primate sleep that can last up to 14 hours. All these animals share the need for sleep, despite huge differences in the length of rest, the frequency and duration of REM or paradoxical sleep, and the choice of bedding and posture.

Mammals prevented from sleeping get sick and die. And if REM sleep is suppressed in mammals, they suffer health problems. There are exceptions to these fundamental rules and many questions remain unanswered as to why mammals need to sleep so many hours and why they dream.

Despite the work of scientists in examining the sleep of about 150 species of mammals, we still do not fully understand sleep's real function or that of REM sleep and dreams. All mammals share the same sleep pattern: calm, slow wave sleep alternating with agitated, REM sleep repeated in several cycles during each sleep period. Mammals either rest because they have satisfied their need to eat or procreate, or they want to avoid predators. They also rest to conserve energy or to wait until conditions

Defense posture
Foxes and humans are the main enemies of hedgehogs. To defend itself, the hedgehog rolls itself into a ball, leaving very little of its body unprotected by spines. It maintains this position even when rolled over (see page 86). It also hibernates rolled up. On a daily basis, it sleeps 10 hours.

Rolled up in a ball
Undisturbed, the hazel dormouse (see below) sleeps 15 hours a day. In the fall, it enters hibernation but wakes up periodically to eat.

improve. But there may be other reasons why all mammals must sleep and perhaps future research will reveal them.

The dozing dormouse

Hardly bigger than a large walnut, the hazel dormouse (*Muscardinus avellanarius*) is famous for its sleepiness. It inhabits a cylindrical nest woven from grass and leaves, often built in a jumble of thorn bushes (see page 87). Snug in its shelter, it sleeps soundly all day long. At sunset, it leaves its nest, washes carefully, and sets off to find nuts, seeds, or small berries. This activity lasts for about two hours. The rest of the night is spent taking long naps, balancing precariously on a branch using only its tail as a counterweight. This shaky position is in fact a clever strategy. At the slightest quiver of the branch the dormouse is thrown off balance, wakes up with a jump and takes flight. At about 3 a.m., it returns home to sleep some more.

In preparation for winter, it eats huge amounts of food. By late October, sleeps long and longer each day until it begins to hibernate. Its body temperature hovers just above the ambient temperature, and its heartbeat and breathing are barely perceptible. Similar to other hibernating mammals, the hazel dormouse burns up reserves from a special layer of brown fat to keep it from freezing. Its state of hibernation is so profound that it can be

rolled on a smooth surface without waking up. Occasionally, for reasons unknown to scientists, it will wake up during the winter and go foraging, but it uses up a lot of energy doing this and if it happens too often, it could die. By April, its hibernation over, the hazel dormouse will have lost half its body weight.

Deep sleepers versus shallow sleepers

True hibernation should not be confused with a state of dormancy or torpor. Strictly speaking, the term "hibernation" should be reserved for animals whose body temperature drops drastically, with reduced heartbeat and breathing. Hibernation always commences when the animal is sleeping. REM or paradoxical sleep diminishes little by little until it disappears completely and gives way to slow wave sleep.

Animals in hibernation do not wake up when they are disturbed whereas animals in dormancy wake up almost instantly. A bear in its winter den comes awake immediately if something pokes its head into its home. A dormant grizzly bear's heart rate will slow down from about 50 beats a minute to 10 or 12 and its temperature will drop only a few degrees. Skunks and raccoons are also shallow sleepers who just lower their body temperature a couple of degrees and wake up periodically to forage. Then there are those who are dormant on a daily basis such as pygmy mice (and hummingbirds) who must use this strategy to survive.

An example of a true hibernator is the groundhog (also called woodchuck or marmot). This rodent digs an elaborate burrow in the spring with rooms for sleeping, eating and eliminating. All summer long, it builds up a store of fat under its fur and stores away food in its pantry. In the summer, it loves to bask in the sun remaining motionless for long periods of time, lying on its side near the entrance to its burrow with its head resting on its forelegs. From a distance it seems asleep. In fact, it is very much on the alert, resting with its eyes open. If an eagle or hawk is spied on the horizon, a sharp whistling cry goes up to warn other groundhogs and everyone beats a retreat to their burrows. As the weather gets colder and the days get shorter, it heads for its bedroom where it will fall into a deep sleep on a soft pile of grass and leaves. It curls up into a ball and its heart slows from 80 beats

Yawning is contagious
From the young orangutan (page 90) to the cheetah (above) and the domestic cat (below), all mammals yawn when they are sleepy. Cats sleep more than 15 hours a day, in catnaps of varying length. Primates tend to sleep without interruption through the night but they are not averse to naps during the day.

91

Sleeping in the sun
Verreaux's sifaka (Propithecus verreauxi) from Madagascar passes the night asleep on a tree branch, safe from predators, but it loves to bask in the sun during the day. It eats fruit, leaves and buds and can make prodigious leaps of 10 meters (33 ft).

Almost asleep
The red panda (see page 93) is nocturnal and arboreal, eating bamboo, grasses, fruit, buds, leaves and acorns. It sleeps during the day in a tree, usually curled up with its bushy tail over its head or, like a raccoon, with its head tucked below its chest and between its forelegs. The giant panda (see page 95) is also nocturnal but spends most of its time on the ground, occasionally climbing trees.

a minute to about 4 or 5 beats. Its body temperature drops from 38°C (100°F) to about 7°C (45°F). It is in true hibernation; nevertheless, it wakes up every four weeks or so to nibble on some food and to use its toilet. Sometimes, one can be seen on the snow, looking around vaguely, before returning to its sleeping chamber.

Another deep sleeper is the ground squirrel. The Arctic ground squirrel only spends four or five months of the year awake. The rest of the time it hibernates in its well-stocked burrow with its body temperature reduced from 32.2°C (90°F) to 4.4°C (40°F) and its heart rate reduced from 150 to 5 beats a minute. At the same time in the hot deserts of the American southwest, ground squirrels are estivating, deep in sleep until the weather brings rain and with it new plant life.

In an experiment to find out more about hibernation, a scientist injected some blood from a hibernating ground squirrel into non-hibernating squirrels. Shortly after, the non-hibernators went into hibernation. Clearly, there is a hormonal factor that triggers the long sleep.

How do hibernating mammals survive the winter? Along with regular white fat, they have special brown fat in patches across their shoulders and back which delivers quick energy when it is needed. To wake up, the brain has to warm up first to give the necessary orders to the rest of the body. As the brown fat is close to the brain, heart and lungs, it delivers the energy to these main organs first. When a ground squirrel wakes up from hibernation, its front legs move before the hind legs because the brown fat is farther away from them.

Little sleepers
Because it must eat constantly to keep warm the shrew is a voracious feeder. If it goes without food for more than half a day, it dies. Requiring food almost constantly means that sleep cannot last longer than a few minutes.

The timorous roe deer sleeps only 3 to 4 hours a day. When the female is nursing young, she sleeps only 1.5 hours a day. She never closes her eyes for more than 10 minutes at a time and has such light sleep that the slightest noise or smell will wake her.

92

The elephant sleeps as little as 4 hours a night, half of which are spent standing. The giraffe takes even less time to sleep: 2 hours.

How can these mammals survive on so little sleep? The answer may be that they sleep in short episodes and compensate for their lack of sleep by drowsiness while awake. The Congo okapi sleeps less than its relative, the giraffe: 5 minutes in 24 hours and even those 5 minutes are not in one stretch! It sleeps in 10 to 30-second naps, each of which includes a few seconds of paradoxical sleep.

During REM sleep, muscles relax and the mammals have to lie down. How long and how soundly they sleep is linked to where they choose to bed down. The safer the spot, the better they sleep. To avoid predators the western baboon, which lives on the African savanna, sleeps on its heels in the uppermost branches of trees. In such an uncomfortable position, it looks nearly impossible to sleep soundly—let alone get REM sleep—for fear of falling. Yet, they must get enough sleep or they would not survive.

On land and sea
Elephants (above and pages 96–97) only sleep about 4 hours a day. Sea lions (below) can sleep on land and in the sea. On land, they sleep in a number of positions: on the side, belly or back. In the water, they paddle with one flipper to stay afloat. Like seals, they sleep one hemisphere at a time to avoid drowning.

In the 1980s, Russian biologist Lev Mukhametov solved some of the mystery of sleep in marine mammals. Marine mammals have had to develop a type of sleep specially adapted to life in water and to conscious breathing. Underwater, the seal takes short naps. EEGs carried out have revealed that the seal switches from slow wave sleep while it swims and comes up for air at the surface, to REM sleep while it pauses to breathe. The sea lion basically sleeps one hemisphere at a time with EEGs indicating short periods of slow wave sleep and REM sleep in the sleeping hemisphere. No REM sleep has been detected in the dolphin, which is a mystery. The dolphin sleeps for 7 hours, one hemisphere at a time, but only in slow wave sleep. The other hemisphere remains alert so that the dolphin can continue swimming and breathing.

Posture, bedding and temperature

Mammals have many sleeping positions. The sloth (see page 99) and bat sleep hanging with their heads down. Carnivores often sleep curled in a ball but lie sprawled out if it is hot. A seal floats vertically in the water. The elephant uses its trunk to build a little pillow under its head before going to bed. The giraffe may remain standing up for most of its short sleep but it sometimes leans its head on a branch to ease its neck. When it is getting its REM sleep and lies down, it rests its neck along its back.

Mountain gorillas change their leafy nests every evening, building on the ground or in trees for the younger and lighter members of the troop. This daily bed change makes sure that parasites do not have time to infest the bedding. Similarly, an orangutan (which does not live in a social group) makes its nest for a single night's stay but does so between 18 and 28 meters (60 and 90 ft) up in the trees. A male will test the branches by swinging on them before building his nest to ensure the tree can

A social yawn
Sometimes yawning has social implications. In a herd of elephants, it only needs one elephant to begin yawning for all the others to follow suit. It is a sign that the herd will shortly turn in for the night.

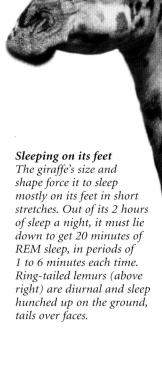

Sleeping on its feet
The giraffe's size and shape force it to sleep mostly on its feet in short stretches. Out of its 2 hours of sleep a night, it must lie down to get 20 minutes of REM sleep, in periods of 1 to 6 minutes each time. Ring-tailed lemurs (above right) are diurnal and sleep hunched up on the ground, tails over faces.

withstand his weight of more than 90 kg (200 lbs) for 14 hours of sleep. Once he has checked to see if it is strong enough, he skillfully weaves a mattress out of vines in a few minutes. A female builds a separate nest for herself and her offspring.

Bedding is not the most important factor for sleep. Sleep depends on the internal biological clock, circadian rhythms, state of tiredness and the length of time the animal has stayed awake, but the safety factor has a part to play, too. If an animal is under stress, it cannot fall asleep. Unless the animal feels safe, the system that keeps it on the alert will not shut down.

Another important factor is body temperature. For an animal to fall asleep, the surroundings must be neither too hot nor too cold. In the cat, temperature is measured by thermal receptors in the ears and on the face. In all the species that an American physiologist, Swann, studied thermal receptors read exactly 27.5°C (81.6°F) when the animal fell asleep. The desert kangaroo rat protects itself against the day's sweltering heat by digging a well-ventilated underground tunnel. By facing the draft of air, and by using its own saliva for evaporation, the heat receptors maintain the same 27.5°C (81.6°F) temperature.

Different types of sleep

Most mammals go through the first stages of slow wave sleep and end up with agitated or REM sleep and then back again through the cycle a number of times in a sleep period. However, some of these periods may be very short. In marine mammals such as dolphins, the REM sleep is missing altogether.

The brain uses neurotransmitters such as histamine, acetylcholine, serotonin, norepinephrine, dopamine, GABA and so on, to maintain and regulate the sleep-wake states. During wakefulness, all cerebral systems are active except those linked to sleep. Slow EEG waves are characteristic of quiet sleep. In agitated sleep, EEG waves are rapid and similar to those produced in an awake state; hence the term paradoxical sleep, coined by Michel Jouvet. Paradoxical sleep or REM sleep is accompanied by a significant increase in heart and respiratory rates. The paradox lies in the high level of brain activity yet complete absence of muscle tone.

While asleep, mammals' muscles are relaxed except for sexual organs and the eyes, which move rapidly beneath closed eyelids. The reasons for eye movements and erection remain a mystery.

In humans, paradoxical sleep is synonymous with dreaming. Faced with the impossible task of getting a dreaming animal to wake up and tell us what it was dreaming about, a method for proving the existence of dreams in animals had to be devised. In the 1960s, Michel Jouvet showed that if brain stem neurons responsible for suppressing muscle movement were destroyed, removing the REM paralysis, a cat would act out what it was dreaming. And that is exactly what happened: the cat began to pursue its prey, then defended itself against an enemy, and washed itself as it would in the awake state.

Sleep is regulated by circadian rhythms, which date back to the

Slothful behavior
The sloth lives its life hanging from the branches of trees, either sleeping or dozing 90% of the time. It moves in slow motion, leaving its tree once a week to defecate. The sloth moves so little that green algae set up residence in its fur, giving it a greenish hue.

origin of life. The hamster, for example, has clear-cut, well-defined cycles, sleeping during the day and being active at night. Studies on this rodent have allowed scientists to locate the body clock in mammals within a group of neurons in the hypothalamus. Although this body clock may be responsible for synchronizing sleep, hormones are actually what triggers sleep. Today, studies in sleep genetics are aimed at discovering what genes are involved in different sleep mechanisms. Scientists are trying to isolate the genes responsible for sleep in mice and in particular those governing sleep duration and REM sleep. Experiments carried out on mice that were prevented from getting REM sleep show a marked increase in paradoxical sleep when later allowed to rest, thus proving that REM sleep is indeed a basic requirement.

Paradoxical sleep

Paradoxical or REM sleep probably appeared about 180 million years ago. This date is supported by the opossum—a "living fossil"—which has been around that long and exhibits REM sleep. Each species has its own periods of paradoxical sleep of differing frequency and duration. In mice they occur every 4 minutes; every 5 minutes in the chinchilla or bat, every 12 minutes in the squirrel, pig or sheep. The frequency of these dream periods increases to every 20 minutes in the mole, pony, cow, opossum, and fox; every 30 minutes in felines; every 40 minutes in macaques or baboons; every 60 minutes in horses or tapirs; and in large primates such as humans, every 90 minutes. In fact, the frequency of paradoxical sleep is linked to

Predators sleep soundly
In mammals, predators sleep longer hours than their prey. Animals such as the hare or antelope are restless sleepers, always on the alert, snatching fitful sleep when they can, unable to enjoy long stretches of unbroken rest. This snow leopard (above) hunts in early morning and late afternoon hours but will nap in between, saving the night for its soundest sleep.

Sleeping in one long stretch
As a result of the evolutionary process, primates developed a system of unbroken sleep. Chimpanzees, gorillas and orangutans sleep in one long stretch, usually at night, just like humans. Chimps and baboons sleep about 10 hours, gorillas for 12 hours, and orangutans about 14 hours.

metabolism. A mouse, with its higher metabolism, will have more dreams than an elephant, but a mouse's individual dreams will be shorter in length.

However, irrespective of the species and sleep duration, each sleep cycle is made up of 75% slow wave sleep followed by 25% REM sleep.

Overall, predators enjoy a deeper, longer sleep filled with more dreams than their prey. With the exception of the dolphin, the echidna and the duck-billed platypus, all mammals share the same alternating pattern of slow wave–fast wave sleep. The giant armadillo with a daily dose of 6 hours of paradoxical sleep and the opossum, a close runner-up with 5 hours, are champion dreamers.

Age and level of development at birth also influence the amount of paradoxical sleep. Adult mammals have five times less paradoxical sleep than their young (this also holds true for birds). The less developed an animal at birth, the greater the amount of paradoxical sleep. Mammals born immature—and even

102

premature in man—have only paradoxical sleep. Slow wave sleep develops later.

Kittens are born completely dependent on their mother and physically immature. For the first 10 days of life, a kitten's sleep is 90% paradoxical. The foal, on the other hand, which can frolic and graze within hours of being born, has only a very small proportion of paradoxical sleep.

Why do we sleep?

The brain seems to be the only organ to suffer from a lack of sleep. Since the 1970s, Dr. Allan Rechtschaffen's research team at the University of Chicago has been studying the effects of sleep deprivation in rats. Under laboratory conditions, rats completely deprived of sleep died after 13 to 21 days. At the start of the experiment, an increase in the animal's body temperature was recorded. In the last stages, rats steadily lost weight and body temperature dropped. If the rats were allowed to sleep before they

Primate beds
Both orangutans (above left) and gorillas (above and pages 104–105) use branches and leaves to make nests in the trees for sleeping at night. Gorillas also make "day nests" for napping, and these are often on the ground. They usually use the nests for only one night before moving on the next day to forage elsewhere. Orangutans are solitary creatures but gorillas live in a social group so their nests are close together.

got to the stage where their body temperature dropped, they recuperated quickly by sleeping longer than usual, without any organ suffering permanent damage. Carol Eversen, from Memphis University, maintains that death through lack of sleep is due to a collapse of the immune systems. However, University of Chicago research scientists have shown that giving antibiotics does not prevent the rat's death. Others maintain that death is caused more by the stress endured during the experiment than by lack of sleep. When autopsies were done on the dead rats, they found nothing wrong with any of the organs, no infections, no vitamin deficiencies.

Jim Horne, a lecturer at the Sleep Research Center at Loughborough University (Great Britain) has investigated the importance of slow wave sleep in brain recuperation. During recovery from sleep deprivation, slow wave sleep takes priority over all other forms of sleep. Contrary to paradoxical sleep, slow wave sleep is

Domesticated sleep
A pet cat will stay awake during the day because it is imitating its human masters. Cats in the wild usually hunt at night. The domesticated cat has swapped rhythms of sleeping and waking, but if left alone during the day, a well-fed cat will have long drawn-out naps, sometimes resulting in wakeful hours during the night. It still manages to sleep for 15 hours a day.

vital for neuron synchrony. On an EEG the waves specific to slow wave sleep are particularly visible in the prefrontal cortex. According to Horne, sleep has undergone change in mammals: "Seeing as the prefrontal cortex is underdeveloped in the rat, one would suppose that the function of slow wave sleep has been modified during evolution in mammals. Primarily because it is a factor in immobility and energy saving, it reaches its culmination in primates, and man in particular, by facilitating brain recuperation functions."

Not everyone agrees about the role of slow wave sleep in brain repair, particularly Russian biologist Lev Mukhametov, a respected specialist in marine mammals' sleep. He says that the dolphin is a prime example of how sleep's essential function does not lie in muscle rest and organ repair because the dolphin moves continually while asleep. According to Mukhametov, there is no doubt that repair of the brain's neural network is the answer to questions about animals' sleep.

Why do we dream?

As for paradoxical or REM sleep, apart from the role it plays in the maturation of the neonatal brain, it remains a mystery. Even if it's not vital, it must have a function or mammals would not have evolved this system or retained it for so long.

Recent theories link paradoxical sleep to the past and to memory. Is it used for purging memory? Michel Jouvet's research centers around paradoxical sleep's use for developing neural readiness, introducing a concept of programming. Paradoxical sleep would enable the brain to be stimulated and be more efficient upon waking. Above all, it would strengthen individual identity by maintaining and repairing neuron circuits, ensuring that individuality is enhanced. But this remains to be proven.

The dolphin is considered extremely intelligent yet it does not have paradoxical sleep. Dolphins sleep about 7 hours a day, showing only slow wave sleep. The echidna, an egg-laying mammal that combines the features of reptiles and mammals, gets 12 hours a day of slow wave sleep. At the other extreme, the duck-billed platypus, a semi-aquatic, egg-laying mammal, spends 60% of its sleep time in REM-type sleep—more than has been

Upstairs bedroom
The gray fox is a forest animal but can also be found in deserts. A nocturnal hunter, it usually sleeps in burrows or in hollows under rocks or tree roots. As it can climb trees, it will also den in hollow trees. A cave or rock crevice provides a cool place to nap during the hot day. Foxes sleep about 11 hours a day.

Sleeping on the ice
How does the arctic fox manage to sleep in an icy wind? Using the heat-retaining properties of its fur, it buries its snout between its forelegs and covers its face with its tail. Warm breath is enough to maintain heat receptors at 27.5°C (81.6°F), the ideal temperature for dropping off to sleep.

seen in any other animal. What are the dolphins and echidna missing by not having paradoxical sleep?

Finally, one of the patients studied at the Technion Sleep Laboratory by Dr. Peretz Lavie was a man who had sustained brain injuries from shrapnel. To the amazement of the researchers, this man showed no sign of any REM sleep and although he only slept 4.5 to 5 hours a night, he was neither sleepy during the day nor troubled in any emotional way from the lack of REM sleep. He had a wonderful memory and had successfully studied to be a lawyer after his injury. What they did discover was a piece of shrapnel in the part of the brain stem that controls the activation of REM sleep. In his book *The Enchanted World of Sleep*, Dr. Lavie wonders why this system of REM sleep would have evolved and stayed so long if we didn't actually need it: "Could the role played by REM sleep be less vital than we tend to believe? Perhaps it is simply a 'fossilized' relic of brain activity that was necessary in the early stages of our development, somewhere in the transition from cold-blooded reptiles to birds and mammals?"

So, why do mammals dream? To solve this mystery, scientists have days and nights of fascinating study ahead of them.

A mattress made of snow
The polar bear's warm fur (page 108) helps it resist icy conditions. It is so well-insulated that its body heat will not dissipate into the snow. The hollow hairs trap the warmth of the sun and the extra fat helps to generate internal heat. Mother and offspring often sleep in the open but the cub always uses its mother for a pillow. During the bear's dormant period in the winter, the mother will make a den under the snow and sleep away the long, dark hours with her cub.

A nocturnal bear
The Malayan bear or sun bear is the smallest bear in the world. An active climber, it spends most of its day sleeping or dozing in the sun, high up in a tree, in a nest of branches and twigs or in a hollow tree trunk. At nightfall it forages on the ground for rodents, lizards or insects and climbs up trees for fruit to eat. It does not hibernate.

Bibliography

All the Birds of North America, American Bird Conservancy's Field Guide, Jack L. Griggs, HarperPerennial, 1997

Amphibians, Edward R. Ricciuti, Blackbirch Press, 1993

Bird Migration, Robert Burton, Aurum Press Limited, London, 1992

Chameleons, Claudia Schnieper, Carolrhoda Books, 1989

Discovering Crickets and Grasshoppers, Keith Porter, Wayland Publishers, 1986

Do Not Disturb: The Mysteries of Animal Hibernation and Sleep, Margery Facklam, Little, Brown and Company, 1989

Enchanted World of Sleep, Peretz Lavie, Yale University Press, 1996

Encyclopedia of the Animal World, David M. Burn, Peerage Books, 1991

Encyclopedia of Insects & Arachnids, Maurice & Robert Burton, BPC Publishing, 1984

Encyclopedia of Reptiles & Amphibians, Harold G. Cogger & Richard G. Zweifel, American Museum of Natural History, 1998

Encyclopedia of Reptiles, Amphibians & Other Cold-Blooded Animals, Maurice & Robert Burton, BPC Publishing, 1975

Eyewitness Books: Insect, Laurence Mound, Alfred A. Knopf, 1990

Firefly Nature Encyclopedia, David Burnie et al., Firefly Books, 1998

Hibernation and the Hypothalamus, Nicholas Mrosovsky, University of Toronto Press, 1971

Hibernation and Torpor in Mammals and Birds, Charles P. Lyman, Academic Press, 1982

Insect Dormancy: An Ecological Perspective, H.V. Danks, Biological Survey of Canada, 1987

Insect Hibernation, Raintree Publishers, 1986

Insects: Life Cycles and the Seasons, John Brackenbury, Blandford, 1994

International Wildlife Encyclopedia, Maurice & Robert Burton, Marshall Cavendish Corporation, 1969

The Life of Birds, David Attenborough, BBC Worldwide Ltd., 1998

Mysteries & Marvels of Insect Life, Jennifer Owen, Usborne Publishing, 1984

Name That Insect, T.R. New, Oxford Press, 1996

New Larousse Encyclopedia of Animal Life, Maurice Burton, Hamlyn Publishing, 1981

Ontario Birds, Chris Fisher, Lone Pine Publishing, 1996

Oxford Companion to Animal Behaviour, David McFarland, Oxford University Press, 1981

Oxford Companion to the Mind, Richard L. Grégory, Oxford University Press, 1987

The Pinnacle of Life : Consciousness and Self-Awareness in Humans and Animals, Derek Denton, Allen and Unwin, Saint Leonards, NSW, 1993

Portrait of the Animal World, Derek Hastings, Todtri Productions, 1997

Pocket Guide to Insects of the Northern Hemisphere, George C. McGavin, Prospero Books, 1998

Reef Life, Andrea & Antonella Ferrari, Firefly Books, 2002

ROM Field Guide to Amphibians and Reptiles of Ontario, Ross D. MacCulloch, McClelland & Stewart, 2002

Sharks, Andrea & Antonella Ferrari, Firefly Books, 2002

Simple Animals, John Stidworthy, Facts on File, 1990

Sleep Thieves: An Eye-opening Exploration into the Science & Mysteries of Sleep, Stanley Coren, The Free Press, 1996

Snakes of the World, Peter Brazaitis & Myrna E. Watanabe, Michael Friedman Publishing, 1992

Turtles & Tortoises, Vincenzo Ferri, Firefly Books, 2002

Winter-Sleepers, Phyllis Sarasy, Prentice-Hall, 1969

Photographic credits

Index

Page numbers of photographs are in *italics*.